Sports Poems

Compiled by John Foster

OXFORD

Oxford University Press, Great Clarendon St, Oxford, OX2 6DP

Oxford New York
Athens Auckland Bangkok Bogota Bombay
Buenos Aires Calcutta Cape Town Dar es Salaam
Delhi Florence Hong Kong Istanbul Karachi
Kuala Lumpur Madras Madrid Melbourne
Mexico City Nairobi Paris Singapore
Taipei Tokyo Toronto

and associated companies in
Berlin Ibadan

Oxford is a trade mark of Oxford University Press

© Oxford University Press 1991
Reprinted 1992, 1997
Printed in Hong Kong
ISBN 0 19 916428 2
A CIP catalogue record for this book is available from the British Library.

Acknowledgements
The Editor and Publisher wish to thank the following who have kindly given permission for the use of copyright material:

Finola Akister for 'Egg and Spoon Race' © 1990 Finola Akister; Mary Dawson for 'Obstacle Race' © 1990 Mary Dawson; Michael Glover for 'Racing the Wind' © 1990 Michael Glover; Theresa Heine for 'Sports Day' © 1990 Theresa Heine; Brian Moses for 'The Wheelchair Race' © 1990 Brian Moses; Judith Nicholls for 'Sack Race' © 1990 Judith Nicholls; Irene Yates for 'The Flying Reptiles Race' and 'The Fastest Runner' © 1990 Irene Yates.

Although every effort has been made to contact the owners of copyright material, a few have been impossible to trace, but if they contact the Publisher correct acknowledgement will be made in future editions.

Illustrations by Bucket, Norman Johnson, Peet Ellison, David Parkins, Alex Brychta, Joe Wright

Obstacle Race

Over the tree trunks, under the net,
Crossing the stream without getting wet;
Over the planks on their hands and knees,
And wriggling through tyres that hang from trees.
Then running uphill to the end of the track;
And a prize for the team that's first to get back.

Mary Dawson

Sack Race

Toes in,
knees in.
Quick now,
squeeze in!
Itchy back,
tickle-knees,
hairy sack
makes you sneeze.
Two-feet-hop,
never stop!
Snap, snip,
don't trip . . .
There and back
jumping sack . . .
One . . .
    two . . .
        three . . .
           OFF!

Judith Nicholls

Racing the Wind

I said to the wind –
'I'll race you then
To that gate there
And back again!'
But the wind said to me –
'How will we tell
Which of us won?
I'm invisible!'
So I thought and thought,
And then I found
A crisp bag
Crumpled on the ground.
I picked it up
And said to the wind –
'You blow *this*
And I'll just run . . .'

So off we went,
Me and that bag,
Dashing like mad,
Not once looking back.
But when we got there
The bag went on,
Bouncing, flying
Right across
The playing fields
Till it got lost.
Well, I raced back
To the starting line,
Shouting, 'I've won!
Your bag can't run
In a straight line!'
The wind just didn't know
What to say.
It huffed and puffed
And gruffed around,
Rattling the chimneypots
All day,
Blowing across and round and down,
Searching for the bag it lost . . .

Michael Glover

The Flying Reptiles Race

Five flying reptiles were just about to dine.
The dinner had arrived and it looked just fine.
Then up jumped a bossy one and shouted with glee,
'I bet that I could beat you to the Far-Away Tree!'

The other reptiles laughed and they cried, 'No way!
We're the fastest in the land, we could beat you any day!'
The bossy one boasted, 'I am the fastest one!'
But they all disagreed. So the race was on.

They lined up on the cliff edge ready to begin.
Five flying reptiles each saying, 'I'll win!'
They gazed across the ocean stretching far beyond the sand.
'The winner,' said the bossy one, 'is first back to land.'

Then 'Go!' screeched the bossy one giving them a fright –
And four foolish reptiles flew off into the night.
One bossy greedy reptile went off alone to dine,
'They won't be back till dawn,' he said, 'the dinner is all mine!'

Irene Yates

The Wheelchair Race

We were side by side in the corridor,
trying to pass the time,
talking about what we both enjoyed,
his chair parked next to mine.

He showed me how well he whistled.
I told him my drawing was ace.
He asked how fast I could move.
I forget who suggested a race!

He counted us down to zero.
'No dirty tricks,' I said.
We sped along the polished floor.
I made the turn ahead.

We narrowly missed two cleaning ladies,
a nurse and an angry queue,
then knocked a tea trolley sideways,
'Look out, it's the terrible two!'

Our doctor stepped out of his room
to check the dreadful din,
'You might have hurt yourselves,' he said,
but smiled as we wheeled ourselves in.

Brian Moses

Sports Day

My teacher said 'Everyone
Just do your best,
Run as fast as you can
In your shorts and your vest.'

And I ran really fast
Till my legs nearly dropped,
And I reached the white tape
Where they told us to stop.

And my teacher was there,
And she smiled and she said,
'You did run well Peter,'
And she patted my head.

And she reached in a bag
And she gave a rosette
To Thomas MacGregor,
To Paul and to Brett.

And I stood there and waited
For her to reach in-
To the bag, and give me
A rosette with a pin.

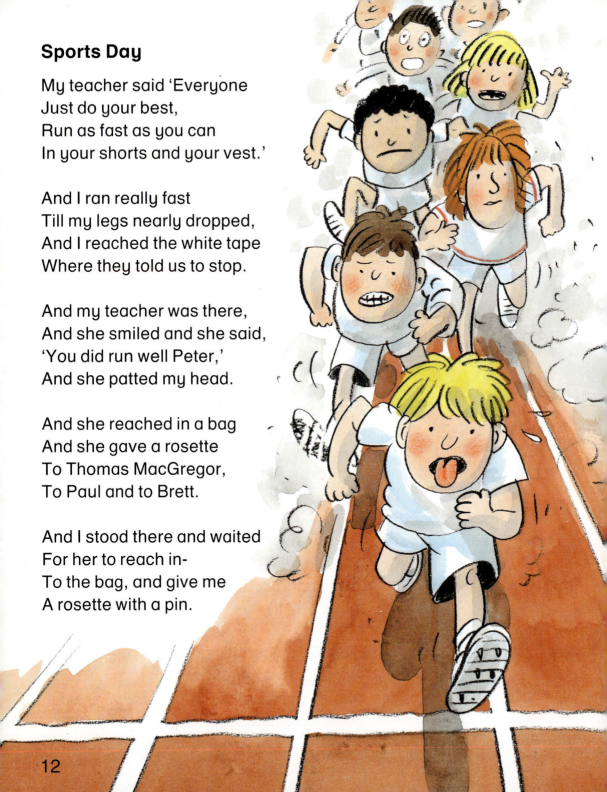

But she just said, 'Please Peter,
Go back to your place,
I have to watch out now
Who wins the next race.'

And I went slowly back,
I had not won a thing
To take home to my Dad,
No rosette with a pin.

But worse than not pinning
A rosette to my chest,
Was to see Wayne O'Rourke
Wearing *two* on his vest!

Theresa Heine

Egg and Spoon Race

When I entered the egg and spoon race,
I knew I was not very quick.
I couldn't run fast so I came in last
And my egg had hatched into a chick.

Finola Akister

The Fastest Runner

Sports day at school
Was ever such fun –
The mums had a race
And who d'you think won?

MY MUM!

Irene Yates

Just when . . .

It's always the same.
Just when you're playing a game,
Just when it's exciting
And interesting
With everyone racing
And chasing,
Just when you're having so much fun,
Somebody always wants something done!

Max Fatchen